FOCUS ON

ELEMENTARY

Biology

Teacher's Manual

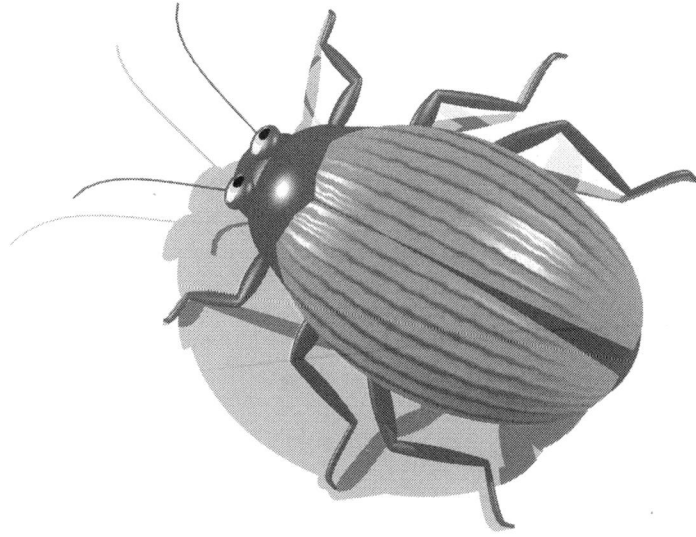

Rebecca W. Keller, PhD

REAL SCIENCE 4 Kids

Cover design: David Keller
Opening page: David Keller, Rebecca W. Keller, PhD

Copyright ©2013 Gravitas Publications, Inc.

All rights reserved. No part of this publication may be reproduced, stored in a retrieval system, or transmitted, in any form or by any means, electronic, mechanical, photocopying, recording, or otherwise, without prior written permission from the publisher.

Focus On Elementary: Biology Teacher's Manual
ISBN 978-1-936114-52-8

Published by Gravitas Publications, Inc.
www.gravitaspublications.com

Printed in United States

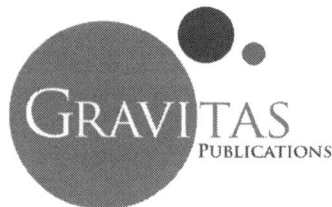

A Note From the Author

This curriculum is designed to provide an introduction to the science of biology for students in kindergarten through fourth grade. The *Focus On Elementary Biology Laboratory Workbook* is intended to be the first step in developing a framework for real science. The series of experiments in the *Laboratory Workbook* will help students develop the skills required for the first step in the scientific method—making good observations. This teacher's manual will help you guide students through the laboratory experiments.

There are different sections in each chapter. The section called *Observe It* helps students explore how to make good observations. In the *Think About It* section, questions are provided for students to think about before they begin the experiment or after they have made their initial observations. In every chapter there is a *What Did You Discover?* section that gives the students an opportunity to summarize the observations they have made. And finally, in each chapter there is a section called *Why?* that provides a short explanation of what students may or may not have observed.

The experiments take up to 1 hour. The materials needed for each experiment are listed on the next two pages as well as at the beginning of each experiment.

Enjoy!

Rebecca W. Keller, PhD

Materials at a Glance

Experiment 1	Experiment 2	Experiment 3	Experiment 4	Experiment 5

cotton balls
rubber ball
tennis ball
banana
apple
rocks
Legos or building
  blocks
other objects
	pencil
	colored pencils
	paper
	
	optional:
	  notebook
		2 small
		  houseplants of
		  the same kind
		  and size
		water
		measuring cup
		
		optional:
		  cardboard box
			2-4 white
			  carnations
			2-3 small jars
			food coloring
			water
			tape
			knife
			colored pencils
				1-2 small clear
				  glass jars
				2 or more dried
				  beans (white,
				  pinto, etc.)
				absorbent white
				  paper
				scissors
				plastic wrap
				clear tape
				1-2 rubber bands
				water

Experiment 6	Experiment 7	Experiment 8	Experiment 9	Experiment 10

microscope with a
  10X objective[1]
plastic microscope
  slides[2]
eye dropper
pond water or
  protozoa kit[3]
	same materials as
	  Experiment 6
	
	optional:
	  baker's yeast
	  Eosin Y stain[4]
	  distilled water
		butterfly kit[5]
		small cage
			tadpole kit[6]
			  (or tadpoles
			  or frog eggs
			  collected locally)
			aquarium
			water
			tadpole food
				clear glass or
				  plastic tank
				plastic wrap
				rubber band or
				  tape
				soil
				small plants
				small bugs, such
				  as:
				    worms
				    ants
				    beetles
				water

Items available from Home Science Tools—www.hometrainingtools.com

[1] student microscope (other sources available online)
[2] plastic microscope slides MS-SLIDSPL or MS-SLPL144
[3] Basic Protozoa Set, LD-PROBASC
[4] Eosin Y stain, CH-EOSIN
[5] Butterfly Garden, LM-BFLYGAR
[6] Grow-a-Frog Kit, LM-GROFROG

Experiments 7 & 8:
Protozoa can also be observed in hay water. To make hay water, cover a clump of dry hay with water, and let it stand for several days at room temperature. Add water as needed.

Materials at a Glance
By type and quantity

Equipment	Materials	Living Things

aquarium for tadpoles
cage, small
cup, measuring
jars, small clear glass (2-3)
knife
microscope with a
  10X objective[1]
scissors
tank, clear glass or plastic
  for ecosystem
	ball, rubber
	ball, tennis
	cotton balls
	eye dropper
	food coloring
	Legos or building blocks
	objects (misc. to sort into
	  groups)
	paper
	paper, absorbent white
	pencil
	pencils, colored
	plastic wrap
	rocks (several)
	rubber bands (1-3)
	slides, plastic microscope[2]
	soil for ecosystem
	tape
	tape, clear
	water
		carnations, white (2-4)
		beans, dried (white, pinto,
		  etc.), 2 or more
		houseplants of the same kind
		  and size, small (2)
		pond water or protozoa kit[3]
		tadpole kit[6]
		  (or tadpoles or frog eggs
		  collected locally)
		
		For ecosystem:
		  plants, small, several
		  ants
		  beetles
		  worms
		
		optional:
		  baker's yeast
		  butterfly kit[5]

Food Items		

apple
banana
tadpole food
	optional:
	  cardboard box
	  distilled water
	  Eosin Y stain[4]
	  notebook	

Contents

Experiment 1:   **Where Does It Go?**   1

Experiment 2:   **What Do You Need?**   7

Experiment 3:   **Who Needs Light?**   12

Experiment 4:   **Thirsty Flowers**   16

Experiment 5:   **Growing Seeds**   20

Experiment 6:   **Little Creatures Move**   24

Experiment 7:   **Little Creatures Eat**   28

Experiment 8:   **Butterflies Flutter By**   32

Experiment 9:   **Tadpoles to Frogs**   35

Experiment 10: **Creatures in the Balance**   38

Experiment 1

Where Does It Go?

Materials needed:

- cotton balls
- rubber ball
- tennis ball
- banana
- apple
- rocks
- Legos or building blocks
- other objects

Objectives

In this unit students will try to sort different items into different groups.

The objectives of this lesson are:

- *To help students make good observations.*
- *To help students understand that there are different ways to sort objects.*
- *To develop a vocabulary to describe the objects that are observed.*

Experiment

I. Observe It

In this section students will make careful observations of each object they have collected.

❶ Help the students collect a variety of objects to be sorted into groups. The Materials List on the previous page gives some suggestions for items to be used.

Have the students put the objects on a table and look carefully at each item, noting various features.

❷ Encourage the students to use both words and pictures to describe each object. Help them observe different details such as size, color, shape, and texture for each item. Use questions to help them describe the object.

- *What color is a cotton ball?*
- *What color is a banana?*
- *What is the shape of a cotton ball?*
- *What is the shape of a rock?*
- *How would you describe the surface of a tennis ball?*

Encourage the students to use as many different describing words as possible for each item. Their answers may look something like this:

cotton ball			
fuzzy	*round*	*soft*	*white*

tennis ball			
fuzzy	*round*	*hard*	*yellow*

rock

hard	gray	smooth	heavy
_____	_____	_____	_____
_____	_____	_____	_____

building block

square	hard	blue	plastic
_____	_____	_____	_____
_____	_____	_____	_____

❸ Guide the students in noticing features that are similar and different between items.

- *Is a rubber ball larger or smaller than a cotton ball?*
- *Is a rubber ball harder or softer than a cotton ball?*
- *Is a rock like a banana? Why or why not?*
- *Is a tennis ball similar to a banana? Why or why not?*

Next, have the students look at the different objects and the different ways they have described the objects. Help them think about how they might group the objects.

❹ Have the students pick five different groups, and help them write the names of the groups in the space provided. Have the students sort the objects they have collected into the different groups. Each object can only go into ONE group.

round	yellow	small	hard	white
tennis ball	banana	rock	block	cotton ball
			rubber ball	apple

II. Think About It

Help the students notice that some items can fit into more than one group. For example, if they picked both "round" and "yellow," a tennis ball can fit into both of these groups. Have the students think about how they might rearrange the groups, picking different items that go into each group. They can re-sort their items into the groups they've already chosen, or they can pick new groups.

There are no "right" answers, so encourage the students to think about all of the different ways they may want to sort the items.

III. What Did You Discover?

The questions can be answered verbally or in writing, depending on the writing ability of the student. With these questions, help the students think about their observations. Again, there are no "right" answers to these questions, and it is important for the students to write down or discuss what they actually observed. Help them write down or describe which objects could fit in more than one group.

IV. Why?

It is important for students to understand that science is often a dynamic endeavor, and the "answers" that science provides can sometimes change. The identification and grouping of living things can be fairly complicated, and determining exactly which group a living thing belongs to is not trivial.

There are different criteria used to group living things. Most living things are first grouped according to the types of cells they have—plant cells, animal cells, bacterial cells, etc. Once the organism is grouped according to cell type, then the scientist looks for other features to use in categorizing the organism.

Experiment 2

What Do You Need?

Materials needed:

- pencil
- colored pencils
- paper

Optional

- notebook

Objectives

In this unit students will explore the different jobs their parents do in the home. They will also explore what it takes to bring a single tool into the house so that a parent can do a job.

The objectives of this lesson are:

- *To have students observe many of the different jobs needed to run a household.*

- *To help students understand how all of the jobs their parents do are connected to other jobs performed by other people.*

- *To help students understand how cells have many parts that perform interrelated functions.*

Experiment

I. Observe It

Have your students follow you (or their parent) around to make observations of jobs performed in the home. Have them bring their workbook with them and record what they see. You may want to have them bring along extra paper or a notebook for recording details. It would be helpful to pick a day where many different jobs can be observed, such as doing laundry, cooking a meal, washing the dishes, repairing a broken door, hanging a picture, cleaning the yard, or mowing the lawn.

Have the students record several jobs they observe being done. Their answers may look something like this:

Job   *cooking breakfast*

Job   *fixing the broken stove*

Job   *planting the garden*

Next, have the students pick one of the jobs on their list. Have them write down all of the items that will be needed for the job to be performed. Their answers may look something like this.

Job _____ *cooking breakfast* _____

Items Needed

eggs	*bowl*
whisk	*frying pan*
butter	*salt*
pepper	*green chili (for NM residents)*

Have the students draw a picture of the selected job being performed and then record the tools used to do the job.

Have the students pick one tool that was used and then draw in detail the item they have selected.

II. Think About It

Next, have the students answer the questions about the item they have selected. They may not be able to answer the questions exactly (they may not know exactly where the bowl or the whisk was purchased), so help them with general answers if needed (a "grocery store" or "hardware store" is enough of a description). Their answers may look something like this:

❶ What is the item?

whisk

❷ Where did the item come from?

grocery store

❸ How did the item get there?

on a truck

❹ Who made the item?

the whisk factory

❺ What is the item made of?

steel

❻ Where does the material that makes the item come from?

from iron ore

If you have time and if the students are interested, it can be helpful to look up some of the ways materials, such as steel, are made. The students will have a better understanding of all the people it takes to produce a product, such as a cooking whisk.

III. What Did You Discover?

Guide the students in answering the questions in this section. Have them think about all of the items people, including their parents, use to do different jobs. Help them think about all of the people it takes to make the items their parents use to do a job.

IV. Why?

This unit helps students understand that when someone does a job, many items are needed for them to be able to do the job, and it actually takes the work of many other people in order to get the job done. This is a good introduction to the concept that many interrelated functions are performed inside a cell to make it operate.

Help the students connect what they observed in this experiment to what goes on inside a cell. In order for someone to have a tool to do a job in a house, that tool needs to come from other people in other cities doing different jobs. Help the students understand that there are "jobs" that proteins do inside cells so that the cells can live. In order for a protein to do its job, it depends on other proteins to do their jobs. For example, for a protein to move molecules from one place to another, other proteins are required to make the molecules that are to be moved.

Help the students see that inside a cell there is a very sophisticated network of proteins and other molecules that do the different jobs that make it possible for the cell to live.

Who Needs Light?

Materials needed:

- 2 small houseplants of the same kind and size
- water
- measuring cup
- cardboard box (optional)

Objectives

In this unit students will observe what happens to a plant if it does not get sunlight.

The objectives of this lesson are:

- *To make careful observations and to compare a plant grown with sunlight to one grown without sunlight.*

- *To introduce the concept of using a "control."*

A "control" is a tool scientists use to compare the specific effect that making a change has on an experiment. By comparing the plant that stays in the sunlight (the control) to a plant that does not get sunlight (the unknown), the students can better observe the effect that the absence of sunlight will have on the plant. Without a control it can be hard to know for certain what caused the observed changes.

Experiment

I. Think About It

❶ Have the students think about what things plants need to have in order to live. Some of the basic things plants need are sunlight, air, water, minerals, etc.

❷ Have the students answer the question that asks what they think will happen if a plant does not get any sunlight. This may seem obvious to the students, but help them think about the details. Use questions such as:

- *What do you think will happen to the leaves if there is no sunlight?*
- *What color do you think the leaves will turn?*
- *What do you think the leaves will feel like after a few days without sunlight? Firm or soft?*
- *How many days do you think it will take for the plant without sunlight to show some problems?*
- *What do you think will happen first?*
- *What do you think will happen last?*

II. Observe It

❶-❷ Have the students look carefully at the two plants.

❸ Help them find words to describe their plants in detail. Have them notice anything different between the plants.

❹ Have them label one plant **A** and the other plant **B**. Have the students draw their plants.

This step sets up the first part of the experiment. It is important to record, in as much detail as possible, the substances and conditions present when an experiment begins. This way, the changes that occur during the experiment can be more easily tracked.

❺-❻ Have the students place the plant labeled **A** in a sunny place and the plant labeled **B** in a dark place. A dark closet would work well, but a cardboard box could also be used as long as it does not let in any light.

❼ Have the students think about what they might observe, and help them write it down.

Have the students water the plants regularly, using the same amount of water for each plant.

Have the students draw what has happened to the plants after one week.

Help them observe any differences. Depending on the type of plant you have selected, it may take several weeks before a significant difference is observed. Have the students observe the plants weekly and record the changes they observe.

III. What Did You Discover?

Have the students answer the questions about what happened to the two plants. Help them write about any significant differences they observed.

IV. Why?

Discuss what happens when a plant does not get enough sunlight to be able to make its own food. Also discuss how using a control helped them compare normal plant growth in sunlight to abnormal plant growth with no sunlight. Help them understand that by using a control, they can make direct comparisons between plants subject to two different conditions—sunlight or no sunlight. Explain to them that a control helped them to determine specifically what effect sunlight, or the lack of it, had on the plants, since the amount of exposure to sunlight was the only factor that was different between the two plants—everything else should have stayed the same.

Experiment 4

Thirsty Flowers

Materials needed:

- 2-4 white carnations
- 2-3 small jars
- food coloring
- water
- tape
- knife
- colored pencils

Objectives

In this unit students will observe how plants get water from the stem to the flower.

The objectives of this lesson are:

- *To make careful observations about how plants use their stems for "drinking" water.*

- *To have students compare what they think will happen to the flower to what they actually observe.*

Experiment

I. Think About It

❶ Have the students think about what will happen to the flower of a carnation if they put the stem in colored water. Help them be as specific as possible. Use questions such as:

- *What do you think will happen to the flower if you put the stem in blue water?*

- *Do you think all of the petals will change color?*

- *Do you think only some of the petals will change color?*

- *Do you think none of the petals will change color?*

- *How do you think the petals may change color? From the end to the center? Or from the center to the end?*

- *Do you think you will be able to see the colored water in the stem?*

- *Do you think the green stem will color the flower?*

- *Will the flower turn green and not blue?*

- *Or if you add yellow will the flower be green and yellow? Or blue and yellow?*

❷ Have the students draw a picture of what they think will happen, showing details

II. Observe It

❶ Have the students carefully observe and draw a white carnation. Help them examine any fine details they find interesting.

❷ Take the carnation and split it in two, lengthwise. Have the students draw the inside of the stem and flower.

❸ Put some water in one of the jars. Add several drops of food coloring, using enough to deeply color the water. You may need to adjust the amounts of water and food coloring. Too much water and too few drops of food coloring will make the dye too dilute, and the coloring of the petals won't be very dark.

Have the students place a carnation in the jar. Make sure the end of the stem is fully submerged in the colored water. You may need to tape the side of the stem to the jar or prop the flower so it does not come out of the water.

Have the students draw the "start" of the experiment. Have them note details, such as how the carnation is fixed to the jar or if it is tilted or how well the stem is submerged.

Help them observe the carnation for the next several minutes. It may take many minutes before the petals of the carnation are fully colored. Have the students make as many observations as possible. Help them pay attention to how the petals are being colored—from the top, side, bottom and so on.

❹ Have the students cut the stem open and observe the inside. Guide their observations with questions such as:

• *Can you see the colored water traveling through the stem?*

• *Can you tell which part of the stem the water is traveling up?*

• *Do you notice anything interesting about the stem?*

III. What Did You Discover?

Have the students answer the questions about what they observed during the colored water and carnation experiment. Help them think about the comparison between the way the carnation looked before and after it was put into the colored water.

IV. Why?

Discuss how the plant "drank" water from the jar. Tell the students that it is similar to how they drink liquid from a straw. When they put their mouth on the straw and suck in the air that is in the straw, liquid moves up from the bottom of their drink. A plant does essentially the same thing, except the "suction" comes from water evaporating from the leaves and petals.

If the students were able to observe differences inside the stem (such as parts of the stem that were light green, dark green, or white), explain to them that a stem has several different types of tissues. One of those tissues (called the xylem) draws water and nutrients up from the soil. Another type of tissue (called the phloem) pulls food back down through the stem to the roots and other parts of the plant. Tell them that only one type of tissue inside the stem draws the water up from the bottom of the jar, and the liquid will not drain back out again.

Experiment 5

Growing Seeds

Materials needed:

- 1-2 small clear glass jars
- 2 or more dried beans (white, pinto, etc.)
- absorbent white paper
- scissors
- plastic wrap
- clear tape
- 1-2 rubber bands
- water

Objectives

In this unit students will observe how a seed grows into a plant.

The objectives of this lesson are:

- *To make careful observations about how a seed grows.*

- *For students to compare what they think will happen to what they actually observe.*

Experiment

I. Think About It

❶ Have the students think about what will happen if they put a bean in a jar, add water, and let it sit for several days. Help them be as specific as possible. Direct their inquiry with questions such as:

- *What do you think will happen to a bean placed in water?*

- *Do you think the roots will come out first?*

- *Do you think the leaves will come out first?*

- *Do you think the bean will change color?*

- *What do you think might happen to the skin on the bean?*

- *How long do you think it will take for the bean to sprout?*

❷ Have the students draw what they think will happen to the bean if they place it in a jar with water and let it sit for several days.

II. Observe It

❶ Have the students carefully observe and draw the bean.
Help them examine any fine details they find interesting.

❷ Take the bean and split it lengthwise into two parts.
Have students draw the inside of the bean with any details they notice.

❸ Help the students carry out the following steps.

Place one or two beans inside and against the side of a jar.

Cut out a piece of absorbent white paper, and wrap it around the inside of the jar. The paper will hold the bean against the side of the jar. Make sure the bean is not touching the bottom of the jar but is placed 6-12 mm (1/4-1/2 inch) above the bottom. You may need to tape the paper to the jar. Pour some water in the bottom of the jar so that the water contacts the absorbent paper but not the bean. The bean will rot if it is in the water. Place plastic wrap on top of the jar and fasten it with a rubber band to seal the jar and prevent evaporation of the water.

Have the students draw the start of the experiment. Have them note details, such as how the bean is oriented in the jar—up, down, sideways, etc.

Help them observe the bean as it grows into a plant. It may take several weeks for the bean to fully sprout. Have the students observe the bean's growth and record all of their observations. If there is more than one bean, have the students record observations for each one, comparing any similarities and differences between the beans. Have the students check the water level periodically.

Allow the beans to fully sprout. Both the roots and leaves should be clearly visible. Have the students note the direction in which the roots grow and the direction in which the leaves grow.

III. What Did You Discover?

Have the students answer the questions about how the beans grew. Help them think about their observations and write summary statements about what they observed. Have them note whether or not the beans grew as they expected.

IV. Why?

Discuss the observations students made about how the beans grew. They should have noticed that the roots of the bean plant emerged first, followed by the leaves. They should also observe that the roots grew down, toward the Earth, and the leaves grew up, toward the Sun.

Ask them how they think a plant "knows" in which direction to grow the roots and in which direction to grow the leaves. Tell them that plant roots have molecules inside that tell roots to grow downward and that leaves have different molecules telling the leaves to grow upward toward the Sun.

Experiment 6

Little Creatures Move

Materials needed:

- microscope with a 10X objective
- plastic microscope slides
- eye dropper
- pond water or protozoa kit

[Protists (protozoa) can also be observed in hay water. To make hay water, cover a clump of dry hay with water, and let it stand for several days at room temperature. Add water as needed.]

The following materials are available from Home Science Tools, www.hometrainingtools.com:

    student microscope (or find online from
        another source)
    plastic microscope slides, MS-SLIDSPL
        or MS-SLPL144
    Basic Protozoa Set, LD-PROBASC

Objectives

In this unit students will look at pond water, hay water, or a protozoa kit to observe how protists (protozoa) move.

The objectives of this lesson are:

- *To make careful observations of protists moving.*

- *To practice using a microscope.*

A microscope that is small and easy for young children to handle is recommended for this experiment. You may need to help your students learn how to look through a microscope lens. For practice, it might help to have the students look at larger objects, such as a piece of paper with lettering they can see. This will help the students orient their eyes for observing small things through the eyepiece. Before beginning the experiment, let them play with the microscope until they are comfortable using it.

Experiment

I. Think About It

The students have read about how protists move. Now have them think about what movement for a protist might look like and what looking at pond water through a microscope might show. Help them explore their ideas with questions such as:

- *What do you think pond water looks like?*

- *Will you see moving creatures?*

- *Do you think you will be able to tell if they are moving? How?*

- *Do you think you will see them rolling or twisting?*

- *Do you think they will swim fast or slow? Straight or in a circle?*

- *What else do you think you might observe in pond water?*

Have them draw what they think they will see when they look at pond water through a microscope.

II. Observe It

This is mainly an observational experiment.

❶ a) Help the students set up the microscope. It helps to place the microscope on a flat, firm surface.

b) Help students put a drop of protozoa water (or pond water or hay water) on a plastic slide.

c) Help the students carefully place the slide in the microscope.

d) Help the students look through the eyepiece at the water on the slide.

It may take several tries before protists can be observed. Help students repeat setting up the slide with samples as many times as necessary.

It is important for students to practice observing as many different details as possible. Help them draw their observations.

❷-❺ There are several drawing frames in the *Laboratory Workbook* for students to fill in with drawings of the different features they observe in the pond water. Encourage them to spend plenty of time looking at all the different features they observe. You can encourage them to stay at the microscope by engaging them with questions such as:

• *What kind of protist do you think you are seeing?*

• *Is it moving fast or slowly? Can you see it spin?*

• *How does it stop? Can it move backwards?*

• *Do you see an amoeba?*

• *How fast does an amoeba move?*

❻-❼ Have students compare some of the protists they are observing. They are asked to make comparisons between different protists of the same kind (two paramecia, for example) and protists of different kinds (possibly a paramecium and an amoeba).

III. What Did You Discover?

Have the students answer the questions about the protists they observed. They should have been able to see different protists moving in different ways. Have them explain what their favorite protist was and why. Encourage them to summarize their answers based on their observations. Help them notice any differences between what they thought they would observe and what they actually observed.

IV. Why?

There are many different kinds of protists. Depending on what your students used for protozoa water, they should have been able to observe at least two different kinds of protists.

Protists move like sophisticated little machines. They roll and spin, stop and start, move forward and back up. Explain to the students how remarkable protists are since they are made with only one cell yet can do so many different things.

Little Creatures Eat

Materials needed:

- microscope with a 10X objective lens
- plastic microscope slides
- eye dropper
- pond water or protozoa kit

- optional:
  baker's yeast
  Eosin Y stain
  distilled water

[Protists (protozoa) can also be observed in hay water. To make hay water, cover a clump of dry hay with water and let it stand for several days at room temperature. Add water as needed.]

The following materials are available from Home Science Tools, www.hometrainingtools.com:

- Student Microscope (other sources available online)
- Glass Depression Slides, MS-SLIDCON or MS-SLIDC12
- Basic Protozoa Set, LD-PROBASC
- Eosin Y stain, CH-EOSIN

Objectives

In this unit students will look at pond water, hay water, or water from a protozoa kit to observe how protists (protozoa) eat.

The objectives of this lesson are:

- *To make careful observations of protists eating.*

- *To practice using a microscope.*

In this experiment students will focus on protists that are eating. If pond water or hay water is being used, there should be plenty of food for the protists to eat.

Baker's yeast stained with Eosin Y can be added to any of the kinds of protozoa water. The Eosin Y stained yeast will be ingested by the protists. Once ingested the red stained yeast will turn blue. It may take some time for this observation.

To make baker's yeast and Eosin Y stain:

- Add 5 milliliters (one teaspoon) of dried yeast to 120 milliliters (1/2 cup) of distilled water. Allow it to dissolve. Let the mixture sit for a few minutes, then add one dropper of Eosin Y to one dropper of the baker's yeast solution and let sit for a few minutes.

- Look at a droplet of the mixture under the microscope. You should be able to see individual yeast cells that are stained red.

Experiment

I. Think About It

The students have read about how protists eat. Have them first think about what it might look like for a protist to eat. Help them explore their ideas with questions such as:

- *How do you think a paramecium eats?*

- *Do you think you can watch it eat?*

- *Do you think you can tell if the food is going inside?*

- *How do you think an amoeba eats?*

- *Do you think an amoeba can eat fast moving food? Why or why not?*

- *What else do you think you might see as the protists eat?*

Have the students draw what they think they will observe through the microscope as they watch protists eat.

II. Observe It

❶ a) Help the students place a small droplet of the protozoa solution onto a microscope slide.

b) If using Eosin Y stained baker's yeast, have the students add a droplet of the stained baker's yeast to the protozoa water on the slide.

c) Help the students carefully place the slide in the microscope.

d) Have the students look through the eyepiece at the protozoa water on the slide.

(You may also position the slide correctly in the microscope and then add the liquids to it.)

It is important for students to practice observing as many different details as possible. Have them draw what they observe.

❷-❺ There are several drawing frames in the *Laboratory Workbook* for students to fill in with drawings of their observations of protists eating. Encourage the students to spend plenty of time looking at all the different features they observe. You can encourage them to stay at the microscope by engaging them with questions such as:

- *What kind of protist do you think you are seeing?*

- *Is it eating?*

- *Can you tell what it is eating?*

- *Can you tell if the protist is eating another protist or something else?*

- *How fast does it eat?*

❻-❼ Have the students compare some of the protists they are observing. They can make comparisons between different protists of the same kind (two paramecia, for example) and protists of different kinds (possibly a paramecium and an amoeba).

III. What Did You Discover?

Have students answer the questions about the protists they observed. They should have been able to see different protists eating. Have them explain what their favorite protist was, how it was eating, and why it was their favorite. Encourage them to summarize their answers based on their observations. Help them notice any differences between what they thought they would observe and what they actually observed.

IV. Why?

Different protists eat in different ways. Your students may or may not have been able to observe the protists eating. Explain to them that watching protists eat is sometimes like watching the tigers eat at the zoo. They may not be hungry. Repeat the experiment at a different time if your student was not able to observe protists eating.

Butterflies Flutter By

Materials needed:

- butterfly kit
- small cage

[Butterfly kits can be purchased from a variety of different sources, such as:

Home Science Tools at
www.hometrainingtools.com

Insect Lore at
www.insectlore.com]

Objectives

In this unit students will observe a caterpillar turning into a butterfly.

The objectives of this lesson are:

- *To make careful observations of the metamorphosis of a butterfly.*
- *To learn about a life cycle.*

Experiment

I. Think About It

❶ Have the students think about how a butterfly got its name. Help them look up the origin of the butterfly name from a library or internet reference.

Have them make a drawing of how they think the butterfly got its name.

❷ Have the students think about how a caterpillar turns into a butterfly. Guide their exploration of their ideas with questions such as:

- *What do you think happens first when a caterpillar changes to a butterfly?*
- *What do you think happens next when a caterpillar changes to a butterfly?*
- *What do you think happens last when a caterpillar changes to a butterfly?*
- *Do you think you can watch the caterpillar in the chrysalis as it changes to a butterfly?*
- *What do you think the chrysalis is made of?*
- *What color do you think the chrysalis will be?*

Have the students make a drawing of how they think a caterpillar turns into a butterfly.

II. Observe It

Help the students:

- Follow the instructions to set up the butterfly kit.

- Make careful observations of the different stages of metamorphosis. If they are starting with a chrysalis rather than eggs, have them use the student textbook as a resource for drawing butterfly eggs, or help them find pictures online.

- Draw the life cycle of the butterfly as they observe it.

- Note any other interesting observations.

III. What Did You Discover?

Have the students answer the questions about the life cycle of a butterfly. Help them write summary statements of what they actually observed. They may have expected something different to happen, but encourage them to record what actually happened—even if the butterflies did not grow or the eggs or chrysalis died.

IV. Why?

Help the students understand that the life cycle of a butterfly is a very amazing process. Explain to them that if scientists did not make careful observations, they would not know that a caterpillar and a butterfly are the same creature and would not know about the life cycle of a butterfly.

Discuss with the students the origin of the name "butterfly."

Have the students discuss some reasons why they think scientists might not always be in agreement and why careful observations are important.

Experiment 9

Tadpoles to Frogs

Materials needed:

- tadpole kit (or tadpoles or
  frog eggs collected locally)
  A tadpole kit can be
  purchased from Home
  Science Tools at
  www.hometrainingtools.com.
- aquarium
- water
- tadpole food

Objectives

In this unit students will observe a tadpole turning into an adult frog.

The objectives of this lesson are:

- *To make careful observations of the metamorphosis of a frog.*
- *To learn about a life cycle.*

Experiment

I. Think About It

❶ Have the students think about how the frog got its name. Help them look up the origin of the name "frog" from a library or internet reference source.

Have the students make a drawing of how they think the frog got its name.

❷ Have the students think about how a tadpole turns into a frog. Help them explore their ideas with questions such as:

- *What do you think happens first when a tadpole changes to a frog?*
- *What do you think happens next when a tadpole changes to a frog?*
- *What do you think happens last when a tadpole changes to a frog?*
- *Do you think you can watch the tadpole as it changes to a frog?*
- *What do you think the tadpole will eat when it becomes a frog?*
- *What color do you think the adult frog will be?*

Have the students draw their idea of what will happen as a tadpole changes into a frog.

II. Observe It

- Guide the students in setting up the tadpole kit. If you have collected tadpoles or frog eggs locally, you can find instructions for their care and feeding on the internet.

- Help the students follow the directions included in the kit or obtained online.

- Have the students observe the growth of a tadpole into a frog. Help them make careful observations and drawings, noting how the tadpole changes.

- For *Observe It* ❶: If students are beginning with tadpoles rather than frog eggs, have them draw a picture of frog eggs using the textbook as a resource, or help them find pictures online.

III. What Did You Discover?

Have the students answer the questions about the life cycle of a frog. Help them write summary statements about what they actually observed. They may have expected something different to happen, but encourage them to record what actually did happen—even if the frogs did not grow, or the eggs or tadpoles died.

IV. Why?

Help the students understand that the life cycle of a frog is a very amazing process. Have a discussion about the meaning of metamorphosis as it applies to the frog life cycle.

Explain to the students that scientists would not know about the life cycle of a frog if they did not make careful observations. In the same way, careful observations by the students showed them that a tadpole and a frog are the same creature. They also could see the changes that take place during metamorphosis.

Creatures in Balance

Materials needed:

- clear glass or plastic tank
- plastic wrap to cover the tank
- rubber band or tape
- soil
- small plants
- small bugs, such as:
  - worms
  - ants
  - beetles
- water

Objectives

In this unit students will set up and observe an ecosystem.

The objectives of this lesson are:

- *To make careful observations about how an ecosystem works.*
- *To learn about how different cycles fit together to create a balanced ecosystem.*

In this experiment, students will observe what happens to a closed ecosystem. They will build a terrarium with plants, small animals, soil, and water, and then they will seal the tank. Our world is a "closed" system because of our atmosphere. The students will make a closed system and observe how difficult it is to keep everything balanced.

Experiment

I. Think About It

Have the students think about what would happen if there were too much water, or too much air, or too much sunlight on Earth. Help them explore their ideas with questions such as:

- *What happens when it rains too much?*
- *Where does the water in a city go?*
- *What do you think would happen if there were too much rain for every city?*
- *What do you think would happen to the plants if summer lasted all year?*
- *What do you think would happen to the plants if winter lasted all year?*
- *What do you think would happen if there were too much air?*
- *What color do you think would happen if there were too little air?*

Have them draw or write about one of their ideas.

II. Observe It

Help the students assemble an ecosystem.

Take the plants, animals, and soil, and place them in a container. Add some water to the soil, and seal the container with plastic wrap if it does not have a tight lid.

Have the students record what happens inside the terrarium over several days.

As they observe the terrarium, have them pay attention to changes in:

- The plants:
  their color, their shape, whether or not they are growing, whether or not they are wilting, etc.

- The texture of the soil:
  is the soil too dry, too wet, does it have decay building up, etc.

- The bugs:
  are they eating; are they healthy; are they able to find enough water; etc.

Keep the cover on the tank as long as possible. If the students notice that the plants and animals are dying, you can remove the cover and "reset" the terrarium. Allow the terrarium to regain equilibrium through several days' exposure to the surrounding atmosphere, rejuvenating the plants, and adding more animals, if necessary.

Replace the plastic wrap or lid, and try again. Have the students note why they had to reset the ecosystem and how many days they were able to keep the cover on the tank before needing to reset it. Also have them notice how long it takes to reset the ecosystem.

III. What Did You Discover?

Have the students answer the questions about what they observed happening inside the tank. Encourage them to discuss all of the problems they encountered with the terrarium. Help them summarize their observations.

IV. Why?

The students should discover that it is very difficult to maintain the health of all the plants and animals inside a closed tank. Explain to them that the Earth is a "closed" system like the terrarium, but on a much larger scale. Explain to them that the Earth is delicately balanced. Help them understand that there are food cycles and water cycles and nitrogen cycles that help living things go through their own life cycles. All of these cycles are connected to each other, and when one cycle goes awry, it affects the larger system.

Explain to them that the Earth does a good job of balancing all the cycles and that humans can contribute to the health of the Earth by being aware of how their activities affect this balance. Help them understand that gaining knowledge of the cycles and an understanding of the Earth's balance is a complicated process; therefore, scientists must make careful observations about the world we live in so that we can contribute in ways that help the Earth's ecosystem to stay balanced.

32485650R10030

Made in the USA
Charleston, SC
19 August 2014